Awesome Kitchen Science Experiments

Homemade Science Experiments For Kids

Copyright © 2021

All rights reserved.

DEDICATION

The author and publisher have provided this e-book to you for your personal use only. You may not make this e-book publicly available in any way. Copyright infringement is against the law. If you believe the copy of this e-book you are reading infringes on the author's copyright, please notify the publisher at: https://us.macmillan.com/piracy

Contents

Introduction 1

Food-Based Science Experiments 2

   1. Make rock candy... 2

   2. Make Invisible Ink.. 5

   3. How to get an egg into a bottle.. 7

   4. Naked egg.. 11

Slime, Putty, and Oobleck Science Experiments 15

   1. Dish Soap Silly Putty 15

   2. Oobleck... 18

   3. Soap Soufflé 22

   4. Some slime recipes to make with kids 27

Nature Science Experiments 31

   1. Bean in a jar.. 31

   2. Cut ice cubes in half like magic.. 33

3. Tornado in a bottle 36

Physics and Physical Science Experiments 42

1. How to make a toy parachute 42

2. Blind spot..44

3. Make a paper windmill 46

3. Lava in a cup 49

Introduction

Searching for kid-friendly science experiments? Don't worry if you never made it past biology: These 63 science experiments for kids are super-easy and a lot of fun to boot, as kids are exposed to a wide variety of scientific concepts. It's a great way to spend quality time together as a family and who knows, mom and dad may end up learning a new thing or two, too.

Besides, children are born scientists. They're always experimenting with something, whether they're throwing a plate of spaghetti on the wall, blowing bubbles in the bathwater, or stacking blocks into an intricate tower only to destroy it in one big swipe. But you can actually do some pretty mind-blowing, hands-on science experiments at home using stuff you probably have lying around the house.

Learn interesting science and technology facts by experimenting with different materials that react in surprising ways. Basic materials can help you perform experiments that are simple, safe and perfect for kids. Enjoy our fun science experiments, make cool projects, show friends & family what you've discovered and most importantly, have fun!

Food-Based Science Experiments

1. Make rock candy

Teach kids how to make rock candy for an edible rainy day activity. Making rock candy is also part science experiment, allowing kids hands-on learning with a few simple ingredients and kitchen tools.

Our easy rock candy recipe lets kids observe the crystallization process firsthand while making some pretty delicious treats. Sugar, water, and few more items found at home are all you need to turn your kitchen into a rock candy laboratory.

Step 1

Gather your ingredients and tools. All you need is water, sugar, a clothespin, a pot for boiling, and a few wooden sticks to grow rock candy crystals in your kitchen! You might pick out a food color dye, too.

Step 2

Bring two cups of water to a boil in a large pot on the stove. Next, stir in four cups of sugar. Boil and continue stirring until sugar appears dissolved. This creates a supersaturated sugar solution. This is also the time to add in any flavor enhancements, such as vanilla or peppermint and so on. Allow the solution to cool for 15-20 minutes.

Step 3

While waiting for the solution to cool, prepare your wooden sticks for growing the rock crystals. Wet the wooden sticks and roll them around in granulated sugar. Make sure you allow the sugared sticks to completely dry before continuing to Step 4. You'll need one stick per jar.

Step 4

Once the sugar solution is cool, add in food coloring to create rock candy of your preferred color. Leave this step out for clear-colored crystals.

Step 5

Pour the cooled solution into a glass jar (or jars) and insert the sugar-covered wooden stick into the center of the glass. Make sure that the stick is not touching any part of the jar. If it does, the candy crystals could get stuck to the bottom or to the sides. You can divide the sugar solution across several smaller jars or use one large mason jar, depending on how many sticks of rock candy you'd like to make.

Once in place, secure the stick in place using a clothespin. Cover the top of the glass with a paper towel. You may have to poke a hole in the paper towel for the wooden stick to poke through.

Step 6

Place the glass in a cool and quiet place. Loud noises and a lot of movement can disturb the crystal making process. Every day, the candy crystals will grow larger. They will reach their maximum growth potential by two weeks. When you have a good amount of rock candy crystals, remove the stick and place it on a sheet of wax paper to dry...before eating!

2. Make Invisible Ink

This low-tech invisible ink science experiment lets kids send secret messages to friends and family. All they'll need is a little lemon juice or milk.

Step 1

Gather your ingredients and tools. For this experiment, you need a piece of paper, a cotton swab, a heat source (a lamp or electric stove works), and milk or lemon.

Step 2

If you are using lemon juice, squeeze your lemon into a glass. You can mix it with a little bit of water. Dip your cotton swab into the milk or lemon juice and start writing your message. Let your message dry completely.

Step 3

Once dry, an adult should hold the sheet of paper over a heat source. We used an electric stovetop. You can also use a lamplight or blow-dryer.

Step 4

As the milk or lemon "ink" heats up, it will oxidize and turn brown. You can try this experiment with other substances such as vinegar, honey, or orange juice.

3. How to get an egg into a bottle

It may seem like an impossible feat to get an egg inside of a milk bottle, but with a little scientific understanding and some common household items, it is totally possible! This is a fun, well-known science experiment.

Step 1: Boiling an egg

- Place an egg in a pot full of water. Put the egg in a pot full of water. Use warm water so it will boil faster.
- Boil the water. Put the pot on the stove set to Medium-High temperature. Let it sit there for 20 minutes while the water boils.
- Peel the egg. Empty the boiling water into the sink being careful not to burn yourself. Use cold water to cool the egg, then peel the shell off.

Step 2: Putting the egg in the bottle

#1 Using matches

- Stand the glass bottle upright with the opening skywards. This is the necessary position to perform this trick. The mouth of the bottle should be small, but still at least half the diameter of the egg (like a milk bottle).

- Carefully light three matches. With extreme caution, drop them into the bottle. Wait a second or two.
- Quickly put the egg onto the bottle's opening, wide end up. Don't wait too long to put the egg on the bottle or the matches will go out and this trick won't work.
- After the matches go out, the egg will be pulled into the bottle. Then, you can amaze your friends with your egg in a bottle.

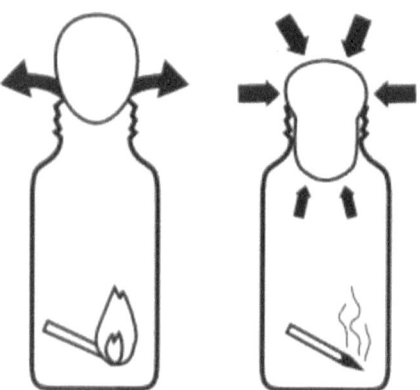

#2 Using Candles

- Use two or three small birthday candles and place them in the narrow end of your peeled hard-boiled egg. Make sure they are securely in place, but not pushed in deep enough to make the egg fall apart.

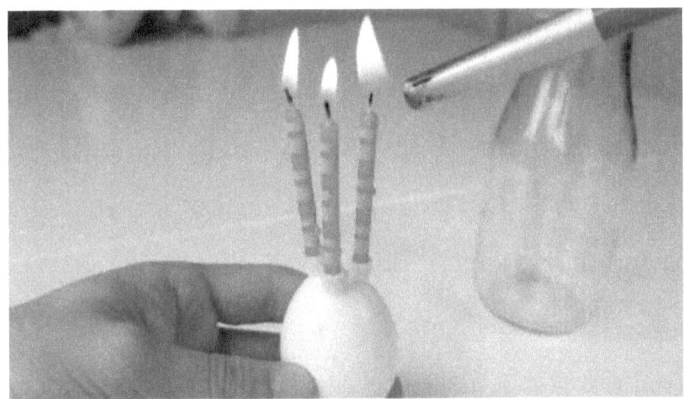

- Carefully (and with adult supervision) light the candles. They should burn easily.
- Take your bottle and place it upside down with the candles inside of it, being careful not to seal the opening with the body of the egg for a few seconds. You need to let the candles warm up the air inside the bottle first.
- After a couple seconds, lower the bottle fully so that the outside opening is sealed by the egg. The candles may go out with a pop, but in moments the egg will slide up into the bottle.

How it works

This trick works because when the matches burn, they heat up the air inside the bottle and release steam (water) as a part of the combustion reaction. This process causes the air inside the bottle to expand, forcing some out of the bottle.

- Once the egg seals the top of the bottle, the matches quickly run out of oxygen and go out. As the air in the bottle cools, the volume of air inside the bottle drops due to condensation of the water vapor (look for the little "cloud" inside the bottle just as the match goes out) and the cooling of dry air.
- When the volume of the air drops, it exerts less pressure on the egg, while the air pressure outside of the bottle doesn't change. The egg is pushed into the bottle once the difference between those forces is sufficient to deform the egg and overcome friction with the neck of the bottle

4. Naked egg

Which came first, the rubber egg or the rubber chicken?

This experiment answers the age-old question, "Which came first, the rubber egg or the rubber chicken?" It's easy to make a rubber, or "naked," egg if you understand the chemistry of removing the hard eggshell. What you're left with is a totally embarrassed, naked egg and a cool piece of science.

Awesome Kitchen Science Experiments

Step 1

Place the egg in a tall glass or jar and cover the egg with vinegar.

Step 2

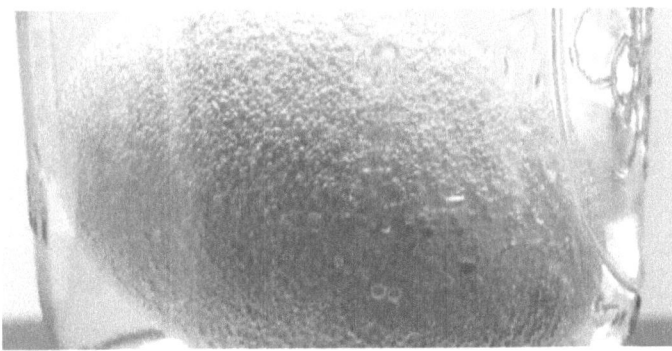

Look closely at the egg. There will likely be tiny bubbles forming on the shell.

Step 3

Leave the egg in the vinegar for a full 24 hours.

Step 4

Change the vinegar on the second day. Carefully pour the old vinegar down the drain and cover the egg with fresh vinegar. Place the glass with the vinegar and egg in a safe place for a week—that's right, 7 days! Don't disturb the egg but pay close attention to the bubbles forming on the surface of the shell (or what's left of it).

Step 5

One week later, pour off the vinegar and carefully rinse the egg with water. The egg looks translucent because the shell is gone! The only thing that remains is a delicate membrane of the egg surrounding the white and the yolk. You've successfully made an egg without a shell. Okay, you didn't really make the egg (the chicken made the egg), you just stripped away the chemical that gives the shell its strength.

How it works

Let's start with the bubbles you saw forming on the shell. The bubbles are carbon dioxide (CO2). Vinegar is an acid called acetic acid (CH3COOH), and white vinegar from the grocery store is usually about 4% acetic acid and 96% water. Eggshells are made up of calcium carbonate (CaCO3). The acetic acid in the vinegar reacts with the calcium carbonate in the eggshell to make calcium acetate plus water and carbon dioxide that you see as bubbles on the surface of the shell.

The chemical reaction looks like this

> **2 CH3COOH + CaCO3 = Ca(CH3COO)2 + H2O + CO2**
>
> Acetic acid + Calcium carbonate = Calcium acetate + Water + Carbon dioxide

The egg looks translucent when you shine a flashlight through it because the hard outside shell is gone. The only part that remains is the thin membrane called a semipermeable membrane.

You might have noticed that the egg got a little bigger after soaking in the vinegar. Here's what happened…Some of the water in the vinegar solution (remember that household vinegar is 96% water) traveled through the egg's membrane in an effort to equalize the concentration of water on both sides of the membrane. This flow of water through a semipermeable membrane is called osmosis.

If you take your naked egg and place it in a glass filled with corn syrup, the egg will shrivel. Since corn syrup has a lower concentration of water than an egg does, the water in the egg moves through the membrane and into the corn syrup to equalize the water concentration levels on both sides.

Slime, Putty, and Oobleck Science Experiments

1. Dish Soap Silly Putty

Ingredients

- 1.5 tablespoons of dish soap and
- 2 tablespoons of corn starch

You already have both of those things, don't you? Well, then you're good to go!

Directions

Step 1

Mix the dish soap and corn starch together as best as you can for about 10 seconds.

Step 2

Once it becomes difficult to stir, get your hands in there!

Step 3

Work the putty until all of the ingredients in the bowl are combined thoroughly. This is when you'll see the putty start to come together.

Awesome Kitchen Science Experiments

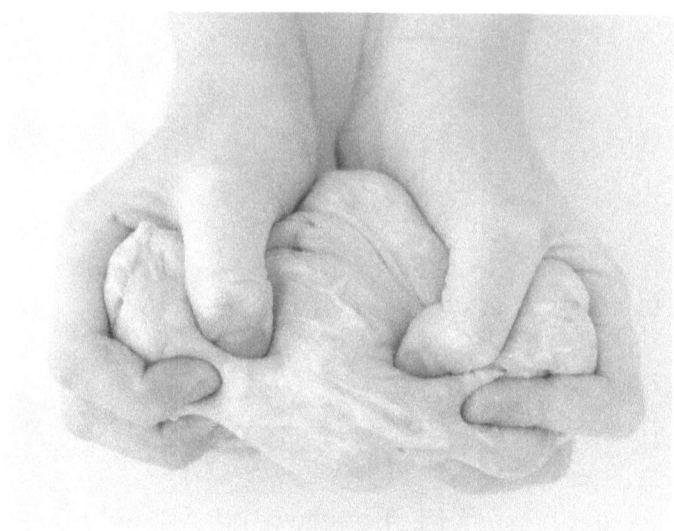

TIP: because dish soap formulas can vary, it's ok if you need to add a little more. If the putty is entirely too dry, add a tiny bit more dish soap to it. Or, if the putty is too runny, add a tiny bit more corn starch to it. After adding more of either ingredient, mix the putty by hand for a few moments. Just a little extra of either ingredient is all it takes.

2. Oobleck

Oobleck is a classic science experiment that's perfect for entertaining both kids and adults. If you haven't seen it in action it's very fascinating stuff and before too long you'll have your hands covered with it, happily making a mess that can be washed away with water.

Oobleck is a non-newtonian fluid. That is, it acts like a liquid when being poured, but like a solid when a force is acting on it. You can grab it and then it will ooze out of your hands. Make enough Oobleck and you can even walk on it!

Oobleck gets its name from the Dr. Seuss book Bartholomew and the Oobleck where a gooey green substance, Oobleck, fell from the sky and wreaked havoc in the kingdom. Here the Oobleck will be made in a bowl and will likely make a mess, but only because you can get carried away playing with it.

Step 1: Materials

All you need is corn starch and food coloring and the food coloring is optional.

Recipe:

- 1 cup water
- 1.5-2 cups corn starch
- A few drops of food coloring of your choice

Step 2: Mix it up

Start with the water in a bowl and start adding the corn starch to it. You can use a spoon at first, but pretty quickly you'll be moving on to using your hand to stir it up.

When you're getting close to adding 1.5 cups of the corn starch, start adding it in more slowly and mixing it in with your hand. The goal is to get a consistency where the Oobleck reaches a state that is the liquid and yet solid.

Sometimes you will need more cornstarch. If so, keep adding more than the initial 1.5 cups. If you add too much, just add some water back into it. You will have to play with it to see what feels appropriately weird.

Step 3: Add food coloring

Now that the Oobleck is just right, it's time to add some color. We save this step for later because it's a fun challenge to stir in the food coloring. You will have to slowly mix the Oobleck around to get it thoroughly mixed.

Step 4: Play With It!

No go ahead and play with the Oobleck. That's the point of all this and you can find lots of tricks to try out. Here's a short list:

- Grab a handful, squeeze it, and let it ooze out your fingers.
- Make a puddle and quickly drag your fingers through it.
- Put it into a plastic container and shake it or quickly bump it against a table.

- Jab at the Oobleck and then slowly let your finger sink in.
- Put it on top of a subwoofer and play some low frequencies at high volume (tough to set up, but worth it)

Have fun and be sure to wash it all off in the end.

How it works

It's all about viscosity, or the liquid's resistance to flow (internal friction), or its thickness. Most fluids are Newtonian (named after Isaac Newton), and will remain at the same viscosity (or rate of flow, or thickness). For example, water has the same resistance/viscosity when standing in a pool as when swimming. It won't change its internal friction or become thicker when you try to move through it.

Oobleck, however, is Non-Newtonian and changes its viscosity. If you apply pressure to the mixture, it increases its viscosity. A quick tap on the surface of Oobleck will make it feel hard, because it forces the cornstarch particles together. Schematic of a fist punching

Oobleck. Long yellow lines are tightly packed together over blue dots to depict how the cornstarch particles get forced together to form a hard surface.

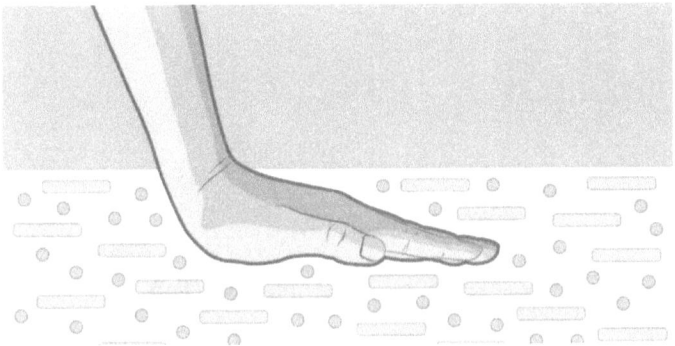

If you dip your hand slowly into the mix, your fingers will slide in as easily as through water. Moving slowly gives the cornstarch particles time to move out of the way. Schematic of a hand gently placed into the Oobleck. Here yellow lines and blue dots are spaced widely apart depicting how the oobleck acts like a fluid in the absence of pressure.

3. Soap Soufflé

Blow up a bar of Ivory soap in the microwave.

Ivory soap . . . it's the soap that floats. But why? Discover the secret behind this floating sensation by cooking the whole bar of soap in the microwave. That's right, a bar of Ivory soap + the microwave oven = a very cool trick! And your kitchen will smell so fresh and clean when you're finished.

The first part of this experiment is designed to prove whether or not the claim is true. Does Ivory soap really float? Fill the bowl with water and drop in a brand-new bar of Ivory soap. It's a pretty simple test . . . does it float?

Maybe all bars of soap float. If you have other brands of soap, try the float or sink test. You'll probably discover that all of the bars of soap sink except for the Ivory brand soap. Why?

Remove the Ivory soap from the water and break it in half to see if the bar of soap is actually hollow or if there are huge pockets of air. If either is true, that would make the soap float, right?

Step 1

Use the knife to cut the bar of Ivory soap into four equal pieces. Place the pieces of soap on a dinner plate, and then place the whole thing in the center of the microwave oven, after asking permission from an adult.

Step 2

Cook the bar of soap on HIGH for 1 minute. Don't take your eyes off the bar of soap as it begins to expand and erupt into beautiful puffy clouds. Be careful not to overcook your soap soufflé.

Step 3

Allow the soap to cool for a minute or so before touching it. Amazing. . . it's puffy but rigid. Don't waste the soap. Take it into the shower or bath. It's still great soap with a slightly diferfent shape and size.

How it works

Ivory soap is one of the few brands of bar soap that floats in water. But when you break the bar of soap into several pieces, there are no large pockets of air inside. If it floats in water and has no large pockets of air, it must mean that the soap itself is less dense than water. Ivory soap floats because air is whipped into the soap during the manufacturing process. If you break the bar of soap in half with your hands and look closely at the edge of the bar, you'll see tiny pockets of air. Cutting the soap with a knife leaves a smooth edge making it impossible to see the exposed air bubbles.

The air-filled soap was actually discovered by accident in 1890 by an employee at Procter & Gamble. While mixing up a batch of soap, the employee forgot to turn off his mixing machine before taking his lunch break. This caused so much air to be whipped into the soap that the batch nearly doubled in size. When the soap was formed into bars, the bars floated in water. The response by the public was so favorable that Procter & Gamble continued to whip air into the soap, capitalizing on the mistake by marketing their new creation as "The Soap that Floats!"

Why does the soap expand in the microwave? This is actually very similar to what happens when popcorn pops or when you try to microwave a marshmallow. Those air bubbles in the soap (or in the popcorn kernels or marshmallow) contain water molecules. Water is also caught up in the matrix of the soap itself. The expanding effect is caused when the water is heated by the microwave. The water vaporizes and the heat causes the trapped air to expand. Likewise, the heat causes the soap itself to soften and become pliable.

This effect is actually a demonstration of Charles's Law. Charles's Law states that as the temperature of a gas increases, so does its volume. When the soap is heated, the molecules of air in the soap move quickly, causing them to move far away from each other. This causes the soap to puff up and expand to an enormous size. Other brands of soap without whipped air tend to melt when heated up in the microwave.

And now the entire kitchen smells like . . . cooked soap.

4. Some slime recipes to make with kids

Your basic slime kit might include the following materials: white school glue, clear glue gel borax, shaving cream, saline solution, re-sealable plastic bags and/or containers, plastic spoons, and bowls. As always, take note of your own children's allergies, and please supervise children when using these materials to ensure everyone's safety.

#1 Basic slime

Step 1

In one of the bowls, mix ¼ cup of water and ¼ cup of white school glue. If you would like colored slime, add a few drops of food coloring to the solution.

Step 2

In another bowl, mix ½ tbs Borax with ½ a cup of water and stir until the borax has dissolved.

Step 3

Add the borax solution to the glue solution and mix. You can also pour the solutions into a re-sealable plastic bag and mix. A slime-like texture should begin to form immediately.

Step 4

Continue to stir and knead until you get the consistency you want.

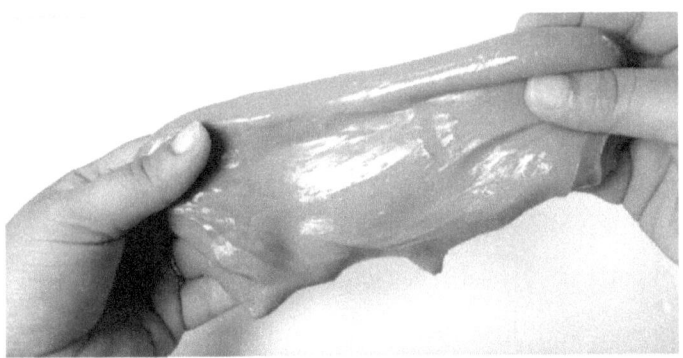

Remove your slime from the bowl (or bag) and enjoy.

#2 Glitter Slime

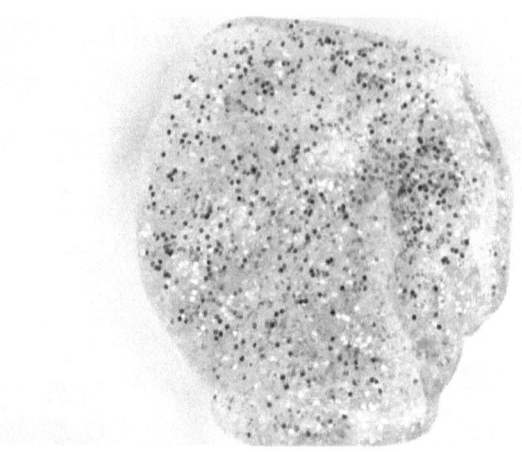

In the Basic Slime Recipe, use clear glue gel instead of white glue and add glitter. The resulting slime is a fun way to play with glitter, without leaving a trail of sparkles around your home.

#3 Found Objects Slime

Add a few pony beads to our re-sealable plastic bag before mixing. Searching for the beads in the otherwise silky slime is very tactilely satisfying. You can even add different sorts of objects and make "Eye Spy" slime!

#4 Fluffy Borax Free Slime

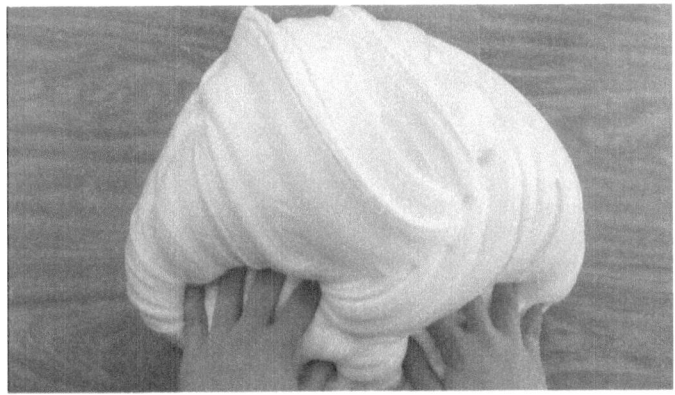

Step 1

Pour about ¼ cup of white glue into a bowl.

Step 2

Add about ½ cup of shaving cream by squirting it into the bowl.

Step 3

With a wooden spoon, mix these two ingredients together until you feel like there isn't any glue left (you can keep squirting small amounts of shaving cream into the bowl to achieve this consistency).

Step 4

Start adding small amounts of contact lens solution to this mixture. Since the contact lens solution is the "slime activator," you can keep adding it until you have fluffy slime.

Nature Science Experiments

1. Bean in a jar

Spring is in the air and it's time to get prepared for blooms and blossoms! A fun project for the Spring season is growing a bean plant in a jar. Guess how fast the plant will grow and what conditions will improve its growth. It's an easy project for the start of the planting season!

Items Needed:

- Jar
- Dry Bean (Lima beans work well)
- Cotton balls or Napkins
- Water
- Sunny Window

Step 1

Wet the cotton balls or napkins and place them in the jar. Cotton balls should be wet but not soggy.

Step 2

Push the cotton balls up against the side of the jar and wedge the bean into the cotton balls so you can watch the bean develop.

Step 3

Put the bean in the jar in a sunny window. Guess what will happen in the next few days. When will it start to grow? How long will it take?

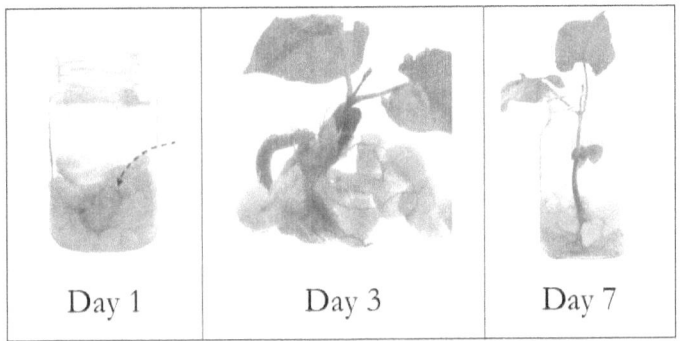

Day 1   Day 3   Day 7

Step 4

Observe the bean over the next few days. Add water to the cotton balls if they start to dry out.

Soon, you'll have a bean plant of your own!

To expand this project and make it more like an experiment, plant multiple beans in separate jars. Put one in a sunny window, one under a lamp, and one in a dark corner. Write down your ideas about what will happen. Measure the plant growth every day for one week to see the differences. Why do you think there are differences?

You can also use three different types of beans – black bean, lima bean, kidney bean – and place them in separate jars, in a sunny window. Measure the plants every day for one week. Which one grows taller? Which one grows fastest? What other differences or similarities do you notice?

When your bean plant outgrows its jar, you can plant it in your garden or in a pot, and take care of it for the season. If keeping it outdoors, make sure the temperature is above 65 degrees Fahrenheit before putting it outside. Add sunshine and water, and it will be a happy plant!

2. Cut ice cubes in half like magic

Step 1

Gather your materials.

Step 2

Cut a piece of wire about 18 inches in length. Twist one end of the wire around a a weight. Twist the other end of the wire around the other weight (If using a soda bottle as a weight, make sure the bottle is completely filled with water.)

Step 3

Place two stools about 6 inches apart from each other. Then, balance the tray across them upside down.

Step 4

Lay the felt on top of the tray to collect the water that accumulates as the ice cube melts. Then, place the ice cube on top of the felt.

Step 5

Place the wire on top of the ice cube so that the weights are dangling on either side of the tray. Guess how long it will take the wire to cut through to the bottom of the ice cube!

Tip: If you want to see the wire as it cuts through the ice, place the wire towards one side of the cube. Don't get too close to the edge though, otherwise your ice cube might fall over!

Watch as the wire magically cuts through the ice cube! Time it! Was your guess correct?

How it works

A weighted wire is going to cut through an ice cube, but somehow leave the ice cube whole after it passes through. This may sound like that magic trick where the assistant gets sawed in half, but there's no magic about this! It's all about the quirky physics of water.

The first quirk of water physics is that, unlike most other solids, ice is less dense than water. When ice gets pressurized, like by a heavy wire pressing down on it, it turns into a denser form. That denser form is liquid water. So when ice gets pressed on too hard, it melts! Bit by bit, the wire melts its way through the ice by turning a small layer into water, sinking through that water, then continuing to press downward.

But when the wire finally gets to the end, you'd hardly know that anything happened — the block of ice will still be solid! This happens because of another quirk of water physics called "regelation." Ice melts into water because of pressure, but once that pressure is gone, any ice above or around it will cool it back down to freezing. Essentially, the ice refreezes itself shut behind the wire. As you watch this experiment happen, see if you can spot any difference between the ice behind the wire and the rest of the ice block.

3. Tornado in a bottle

Create a vortex to drain a bottle of water in seconds!

Step 1

Remove the label from the soda bottle so you have a clear view of the inside. Fill the soda bottle almost to the top with water.

Step 2

Without squeezing the sides of the bottle, turn it over and time how long it takes to empty all of the water. Just hold the bottle upside down. You might want to repeat this several times and average the results. Be sure to use the same amount of water for each trial. Now you're collecting data!

Step 3

Refill the bottle almost to the top with the same amount of water as you did before. When you turn it over this time, move the bottle in a tight, clockwise or counterclockwise circular motion as the water pours out.

Step 4

Keep moving the bottle like this until you see the formation of what looks like a tornado in the bottle. The water begins to swirl, a vortex forms, and water flows out of the bottle very quickly.

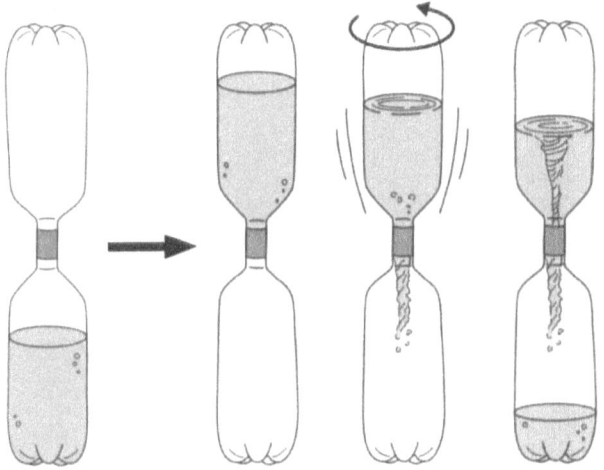

How it works

If you've ever seen a dust devil on a windy day or watched the water drain from the bathtub, you've seen a vortex. A vortex is a type of motion that causes liquids and gases (both are fluids) to travel in spirals around a centerline. A vortex is created when a rotating liquid falls through an opening. Gravity is the force that pulls the liquid into the hole, and the rotation causes a continuous vortex to develop.

Swirling the water in the bottle while pouring it out causes the formation of a vortex that looks like a tornado in the bottle. The formation of the vortex makes it easier for air to come into the bottle and allows the water to pour out faster. If you look carefully, you will be able to see the hole in the middle of the vortex that allows the air to come up inside the bottle. If you do not swirl the water but just allow it to flow out on its own, then the air and water have to take turns passing through the mouth of the bottle.

4. Tinted flower

Do you know that flowers can drink up water and change their own colors into the color of the water they drink?

White flowers are great for doing this color changing experiment. They change colors almost overnight. Other types of flowers such as daisies can take a lot longer (more than 10 days).

Materials:

- White/light-colored flowers such as daisies or white roses
- Food coloring

Tools

- several glasses, vases or test tubes
- adult supervision

Instructions

Step 1

Fill each glass with fresh water from the tap. Put 2-5 drops of food coloring into it, one color each. You can also mix the colors (e.g. blue + yellow = green) to get all the rainbow colors.

Step 2

Trim at least half an inch of stem off the flowers before putting each into the glass and each time you change the water.

Step 3

Add flower food if it is provided.

Step 4

Keep them in a cool place overnight.

Observe the change in colors in the petals.

Try this bonus experiment: cut along the stem into two halves and stop before reaching the flower. Insert each half into a different colored water. Observe how the petals change color.

Physics and Physical Science Experiments

1. How to make a toy parachute

What you'll need

- A plastic bag or light material
- Scissors
- String
- A small object to act as the weight, a little action figure would be perfect

Step 1

Cut out a large square from your plastic bag or material.

Step 2

Trim the edges so it looks like an octagon (an eight sided shape).

Step 3

Cut a small whole near the edge of each side.

Step 4

Attach 8 pieces of string of the same length to each of the holes.

Step 5

Tie the pieces of string to the object you are using as a weight.

Step 6

Use a chair or find a high spot to drop your parachute and test how well it worked, remember that you want it to drop as slow as possible.

Hopefully your parachute will descend slowly to the ground, giving your weight a comfortable landing. When you release the parachute the weight pulls down on the strings and opens up a large surface area of material that uses air resistance to slow it down. The larger the surface area the more air resistance and the slower the parachute will drop.

Cutting a small hole in the middle of the parachute will allow air to slowly pass through it rather than spilling out over one side, this should help the parachute fall straighter.

2. Blind spot

The eye's retina receives and reacts to incoming light and sends signals to the brain, allowing you to see. One part of the retina, however, doesn't give you visual information—this is your eye's "blind spot."

Tools and Materials

- A few 3 × 5 cards or other stiff paper
- Black marking pen (felt tip works best)

Assembly

Mark a dot and a cross on a card as shown.

Step 1

Hold the card at eye level about an arm's length away. Make sure that the cross is on the right.

Step 2

Close your right eye and look directly at the cross with your left eye. Notice that you can also see the dot.

Step 3

Focus on the cross, but be aware of the dot as you slowly bring the card toward your face. The dot will disappear, and then reappear, as you bring the card toward your face. Try moving the card closer and farther to pinpoint exactly where this happens.

Step 4

Now close your left eye and look directly at the dot with your right eye. This time the cross will disappear and reappear as you bring the card slowly toward your face.

How it works

The optic nerve—a bundle of nerve fibers that carries messages from your eye to your brain—passes through one spot on the light-sensitive lining, or retina, of your eye. In this spot, your eye's retina has no light receptors. When you hold the card so the light from the dot falls on this spot, you cannot see the dot. The fovea is an area of the retina that is densely packed with light receptors, giving you the sharpest vision.

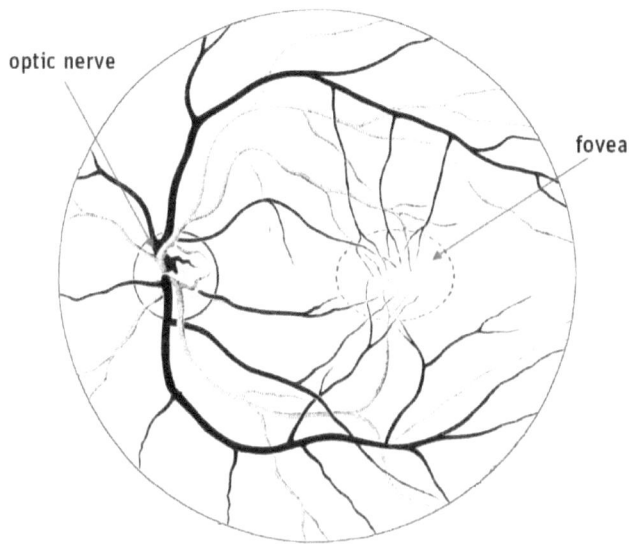

3. Make a paper windmill

You will need:

- Paper in 2 shades
- Decorative stickers
- Flat-ended map pin or push pin
- Short length of thin dowelling (for stick)

Step 1

Cut two 20cm squares of paper, one in each colour. Place one sheet on top of the other. Matching edges all round, fold the paper in half diagonally and open out. Fold diagonally again, this time on the opposite diagonal and open out.

Step 2

Cut from one corner along the diagonal fold stopping around 3cm from the centre. Repeat along the remaining diagonal folds.

Step 3

Fold four alternate corner sections down to the centre of the paper, holding each section under your thumb at the centre as you work. Place a sticker centrally on the windmill so that it holds the four blades in position.

Step 4

Take a map pin or push pin and push it through the center of the sticker and out the back of the windmill. Push the map pin into the top of the dowelling stick so that it firmly secures the windmill in place, but still allows the windmill to turn. You may need to tap gently home with a hammer.

3. Lava in a cup

You will need:

- A clear drinking glass
- 1/4 cup vegetable oil
- 1 teaspoon salt
- Water
- Food coloring (optional)

Step 1

Fill the glass about 3/4 full of water.

Step 2

Add about 5 drops of food coloring – I like red for the lava look.

Step 3

Slowly pour the vegetable oil into the glass. See how the oil floats on top – cool huh? It gets better.

Step 4

Now the fun part: Sprinkle the salt on top of the oil.

Watch blobs of lava move up and down in your glass!

If you liked that, add another teaspoon of salt to keep the effect going.

How it works

So what's going on? Of course, it's not real lava but it does look a bit like a lava lamp your parents may have had. First of all, the oil floats on top of the water because it is lighter than the water. Since the salt is heavier than oil, it sinks down into the water and takes some oil with it, but then the salt dissolves and back up goes the oil! Pretty cool huh?

www.ingramcontent.com/pod-product-compliance
Lightning Source LLC
Chambersburg PA
CBHW030513220526
45464CB00006B/2771